spot

EXTREME WEATHER

FLOODS

by Anastasia Suen

AMICUS | AMICUS INK

boat

sandbags

Look for these words and pictures as you read.

river

car

It rains and rains.
Too much rain will cause a flood.

river

See the river?

Rain fills it up.

Uh-oh. The river overflows!

See the sandbags?
They are stacked.
They try to stop the flood.

sandbags

The water moves fast.

Watch out!

Six inches (15 cm) of water can knock you over.

35
MPH

car

See the car?
It floats.
Two feet (0.6 m) of water
can move a car!

See the boat?

It carries rescuers.

It brings people to safety.

The rain stops.
The water goes down.
Now it is time to clean up.

boat

sandbags

Did you find?

river

car

Spot is published by Amicus and Amicus Ink
P.O. Box 1329, Mankato, MN 56002
www.amicuspublishing.us

Library of Congress Cataloging-in-Publication Data
Names: Suen, Anastasia, author.
Title: Floods / Anastasia Suen.
Description: Mankato, MN : Amicus, [2021] | Series: Spot. Extreme weather | Audience: Ages 4-7. | Audience: Grades K-1.
Identifiers: LCCN 2019036345 | ISBN 9781681519432 (library binding) | ISBN 9781681525907 (paperback) | ISBN 9781645490289 (pdf)
Subjects: LCSH: Floods—Juvenile literature.
Classification: LCC GB1399.S84 2021 | DDC 551.48/9—dc23
LC record available at https://lccn.loc.gov/2019036345

Printed in the United States of America

HC 10 9 8 7 6 5 4 3 2 1
PB 10 9 8 7 6 5 4 3 2 1

Alissa Thielges, editor
Deb Miner, series designer
Ciara Beitlich, book designer & photo researcher

Photos by iStock/kyrien cover, 16; iStock/Andesign101 1; Shutterstock/egd 3; iStock/sanjeri 4–5; iStock/isaravut 6–7; Shutterstock/CEW 8–9; Alamy/michelmond 10–11; Alamy/Adrian Sherratt 12–13; Alamy/Philip Scalia 14–15